NEW VANGUARD 276

GERMAN GUIDED MISSILES OF WORLD WAR II

Fritz-X to Wasserfall and X4

STEVEN J. ZALOGA　　ILLUSTRATED BY JIM LAURIER

OSPREY PUBLISHING

Bloomsbury Publishing Plc

Kemp House, Chawley Park, Cumnor Hill, Oxford OX2 9PH, UK

1385 Broadway, 5th Floor, New York, NY 10018, USA

Email: info@ospreypublishing.com

www.ospreypublishing.com

Osprey Publishing, a division of Bloomsbury Publishing Plc

First published in Great Britain in 2019

A catalog record for this book is available from the British Library.

ISBN: PB 9781472831798; eBook 9781472831941;
ePDF 9781472831934; XML 9781472831927

20 21 22 23 24   10 9 8 7 6 5 4 3 2

Index by Fionbar Lyons
Typeset by PDQ Digital Media Solutions, Bungay, UK
Printed and bound in India by Replika Press Private Ltd.

Osprey Publishing supports the Woodland Trust, the UK's leading woodland conservation charity.

To find out more about our authors and books visit
www.ospreypublishing.com. Here you will find extracts, author interviews, details of forthcoming events and the option to sign up for our newsletter.

Author's note

Unless otherwise noted, all photos here are from official US government sources including the National Archives and Records Administration (NARA II) in College Park, Maryland, and the Ordnance Museum, formerly at Aberdeen Proving Ground, Maryland.

Glossary

AG	Aktiengesellschaft: Share-holder owned company
AVA	Aerodynamische Versuchsanstalt: Aerodynamics Lab
DONAG	Donauländische Apparatebau-Gesellschaft, Vienna
GmbH	Gesellschaft mit beschränkter Haftung; Limited liability company
EK	Lehr- und Erprobungskommando: Training and Testing unit
E-Stelle	Erprobungsstelle; proving ground
FuG	Funk Gerät: Radio device
KG	Kampfgeschwader: Combat wing
OKL-TLR	Oberkommando der Luftwaffe, Technischen Luftrüstung: Luftwaffe High Command, Aviation Technology Development
PC 1400	Panzersprengbombe, Cylindrisch 1,400kg: Cylindrical 1,400kg armor-piercing bomb
RBM	Reichsministerium für Bewaffnung und Munition (Speer's armament ministry)
RLM/GL	Reichsluftministerium Generalluftzugmeister: Reich Air Ministry, Air Force Procurement
SD 1400	Splitterbombe, Dickwandig: Thick-walled 1,400kg fragmentation bomb
STARU	Staßfurter Rundfunk GmbH
WASAG	Westfälisch-Anhaltinische-Sprengstoff-Aktien-Gesellschaft, Reinsdorf
VFK	Vierjahresplan Institut für Kraftfahrzeugforschung

CONTENTS

GERMAN GUIDED MISSILES OF WORLD WAR II

Fritz-X to Wasserfall and X4

INTRODUCTION

World War II German missiles such as the FZG-76 (V-1) and A-4 (V-2) are extremely well known and have been the subject of earlier books in the New Vanguard series.[1] This book is intended to round out the survey of wartime German missile efforts by examining the many other missile programs. Most German missile programs went no further than engineering drawings or simple prototypes, so the focus of this book is on missiles that actually reached combat service or that were approaching production at the time that the war ended in 1945. The two largest categories of these weapons were antiship missiles and antiaircraft missiles.

ANTISHIP MISSILES

German guided missile development began during the Great War with experimental designs using wire-guidance and gyro-based flight controls. This Siemens Schuckert torpedo glider enabled Zeppelins to attack warships from high altitude. Although tests proved promising, the concept was still in its infancy at the end of the war.

German development of air-to-surface missiles can be traced back to 1917, with early attempts by Siemens to develop a glider fitted with an autopilot or wire guidance to deliver torpedoes into the water from a Zeppelin. This proved beyond the state of the art. Blohm & Voss began work on an air-to-surface missile in 1939 as the BV 143, with first flight trials on January 21, 1941, near the Peenemünde missile development center on the Baltic. The BV 143 used a simple gyro-based autopilot to maintain its course. Development dragged on for several years, but the autopilot never proved sufficient for precision attack.

Development of the Fritz-X Guided Bomb

During the Spanish Civil War in 1936–39, Luftwaffe crews noted the difficulties of striking moving warships with conventional bombs. Since it took 20–30 seconds for a bomb to reach a warship from bombing altitudes, the ship could evade the bomb by abrupt maneuvers. What if the bombardier could steer the bomb during the last seconds of flight?

1 Steven Zaloga, *V-1 Flying Bomb 1942–52: Hitler's infamous "doodlebug"*, Osprey New Vanguard 106, Oxford (2005). Steven Zaloga, *V-2 Ballistic Missile 1942–52*, Osprey New Vanguard 82, Oxford (2003).

The Fritz-X consisted of an armor-piercing 1,400kg bomb with a guidance package added to the rear, along with a set of cruciform wings. This example on display after the war is seen in front of a FZG-76/V-1 cruise missile.

The development of German guided bombs began in 1937 under the direction of Dr Max Kramer of the DVL (Deutsche Versuchsanstalt für Luftfahrt: German Aviation Research Institute) in Berlin-Aldershof. Kramer was a specialist in aerodynamics and airfoil design, especially control surfaces such as ailerons. The principal aim of the program was to develop a weapon that could be dropped from high altitude and have a high probability of penetrating the armored deck of enemy warships.

From a tactical perspective, the requirement was to enable a Staffel (squadron) of nine bombers to have a high probability of sinking a warship where otherwise it would take an entire Geschwader (wing) of nine squadrons to sink such a warship using unguided bombs. To start the program, six 250kg bombs were fitted with a simple cruciform wing and a box-tail with simple radio-controlled surfaces, dubbed the X-0. Tests at the Luftwaffe's proving ground, E-Stelle Rechlin, were successful enough that full-scale development was initiated by the RLM/GL CE7 (air ministry bomb development office) under the supervision of Dr Theodor Benecke. In 1940, the firm Ruhrstahl AG in Brackwede was brought into the program to manage the construction of the test weapons due to their experience in bomb manufacture.

A rear view of the Fritz-X tail showing its complex layout. The spoilers used for steering were contained in the horizontal fin.

The Kramer guided bomb was based around either the PC 1400 or SD 1400 bombs, and so in the guided versions were called PC 1400X or

A cross-sectional drawing of the Fritz-X.

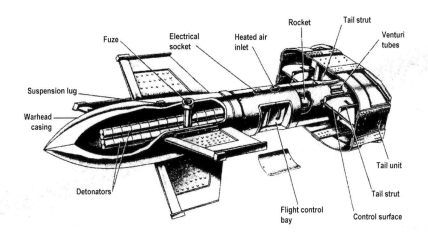

SD 1400X; the "X" referred to the X configuration of the wings. These Luftwaffe 1,400kg bombs were nicknamed "Fritz", so the guided version was widely known as the Fritz-X or generically as the FX 1400.

The Fritz-X could penetrate an armored deck from 130 to 280mm (5–11in) thick depending on its release altitude. The flight control system included two tandem gyroscopes which provided roll stabilization during the bomb's descent. To minimize the power demands of the guidance system, the Fritz-X used a set of spoilers instead of ailerons in the tail section, a feature pioneered by Kramer. These operated in the "bang-bang" fashion, interrupting the airflow over the fins to effect the desired change in directions. "Bang-bang" missile controls are either on or off; there is no intermediate position as there is with conventional control surfaces such as ailerons.

The parent aircraft employed the standard Lofte 7D bomb sight to aim the bomb. Once the bomb was released into the proper trajectory, the bombardier used a Knüppel (club) joystick controller to make any steering corrections necessary for the bomb to strike its target, transmitted to the bomb via a radio link.

The radio command system, codenamed E30 Kehl-Straßburg, was developed under the direction of Ing Theodore Strum of STARU in Staßfurt in coordination with Telefunken GmbH in Berlin. Kehl referred to the Telefunken FuG 203 transmitter located on the aircraft, while Straßburg referred to the FuG 230 radio receiver on the guided bomb. The Fritz-X was fitted with an incandescent flare in the tail to permit daytime tracking by the bombardier, and had a supplementary electric lamp in the tail when dropped at night.

The Fritz-X was intended to have a circular error probability of 5 meters from an altitude of 6,000 meters. Actual performance was highly dependent on the skill of the bombardier and aircraft crew. The first guided tests of the Fritz-X were conducted in February 1942 from an He 111 bomber at the

A   **FRITZ-X ATTACK**
This plate shows a Fritz-X after its release from a Do 217K-2 bomber of KG.100 "Wiking". The weapon had a pyrotechnic flare unit in the tail to assist the bombardier to keep track of the Fritz-X during its descent. The Fritz-X, like the Hs 293, was painted in overall light blue grey (RLM 65 Hellblau) for camouflage purposes.

Baltic test ranges near the Peenemünde missile development center. Due to the fog-shrouded weather conditions over the Baltic, testing was transferred to E-Stelle Sud at Siponto-Foggia, near Naples, in the spring of 1942 by the EK.21 test squadron. During the operational trials, 49 of 100 guided bombs hit their target or came close enough to cause lethal damage.

It was originally planned to operate the Fritz-X from the He 177 Greif heavy bomber, but development problems led to the substitution of the Dornier Do 217K-2 for its initial combat use. The original design of the Fritz-X included 18 preset channels so that a Staffel of nine bombers could launch 18 bombs without radio interference with one another. Although the Do 217K-2 was capable of carrying two Fritz-X bombs, the excessive weight and drag of the weapon led to the practice of carrying a single Fritz-X on one pylon, and an auxiliary drop tank on the other.

The initial production contract was issued to Rheinmetall-Borsig for the first batch of 1,000 bombs at their Marienfelde-Berlin plant. The plans called for a monthly production rate of 300 bombs, ramping up to 750. This production rate was not reached. The program was delayed by problems in the serial manufacture of the Kehl-Straßburg radio control system, and large-scale production did not begin until April 1943. The Fritz-X remained in production through December 1944. There is conflicting data on the number actually built, varying from 1,386 to 2,000.

The most spectacular success of the Fritz-X took place on September 9, 1943, when KG.100 intercepted the three Italian Vittorio Veneto-class battleships off Corsica, sinking the RM *Roma* and damaging the RM *Italia*. The RM *Italia*, previously known as *Littorio*, is seen here a week later on September 16 shortly before reaching the safety of Alexandria harbor under Royal Navy escort.

Development of the Henschel Hs 293 Guided Missile

In 1939, the RLM decided to initiate a guided missile program based around an autopilot flight control system. The program was headed by Prof Herbert Wagner of the Junkers aircraft firm, working in conjunction with Rudolf Brée of the RLM/GL. The DVL encouraged Henschel Flugzeugwerke (HKW) in Berlin-

A preserved example of the Fritz-X on display at the Museum of the US Air Force at Wright Patterson AFB near Dayton, Ohio. (Author)

Schönefeld to set up a research department focusing on novel technologies including jet propulsion and missile guidance. Wagner was hired away from Junkers to lead the new Abteilung F (Forschung: Research), which would become the most important German center for missile development aside from the massive Peenemünde ballistic missile center.

Wagner proposed a long-range, air-launched, glide weapon and the RLM encouraged the use of the Kehl-Straßburg radio command system already in development for the Fritz-X.

The first design was the Hs 293V-1 (V: Versuchs/Experimental) but it never left the drawing board. The first type produced in a small test batch was the Hs 293V-2, built in the winter and spring of 1940. About 100 Hs 293V-2s were assembled. A number of test drops were conducted in May–September 1940. The early versions had repeated guidance failures due to the use of tube (valve) electronics that were extremely prone to failure due to vibration and extreme temperature fluctuations between the airfield and cruising altitude. Wagner gradually substituted more robust electromagnetic relays for many of the tubes. It was also discovered that if hot air was fed into the missile guidance compartment from the parent aircraft, that the tubes were less prone to failure from thermal shock.

Wagner also began to consider a wire guidance system with a pair of wire spools on both the parent aircraft and missile unwinding simultaneously. Wire guidance was viewed as a backup in the event that the radio-command system was jammed by Allied countermeasures.

One of the first lessons from the trials was that an unpowered glider had to be released from high altitude since it took some time before the bombardier could gain control of the missile. This led to greater interest in a powered configuration. The HWK-109-507 rocket motor produced by the Helmuth Walter KG in Kiel was selected. This rocket booster was mounted under the belly of a modified Hs 293V-2 in the summer of 1940, and the first unguided, powered flight was conducted on September 5, 1940, off Peenemünde. An initial guided/powered test drop on December 16, 1940, was unsuccessful when the right/left controls were reversed. The first successful powered/guided drop occurred on December 18 once the mistake was discovered.

The Hs 293V-3 series was the first version regularly built with the Walter rocket booster. In total, about 100 Hs 293V-3s were built, and they were used for further experimental launches in 1941–42 prior to mass production. In November 1940, a special test squadron, later called EK.15, was formed to test the Hs 293 missile.

The Hs 293A-1 was the standard production version of this missile and the second most widely produced German missile of World War II, after the FZG-76/V-1 cruise missile.

The Dornier Do 217E-5 was the most common launch aircraft for the Hs 293 in 1943, such as this one with II./KG.100 at the Istres airbase in southern France in August 1943. Usually, only a single missile was carried, balanced by a 900-liter drop tank on the opposite pylon.

The Kehl-Straßburg guidance system used a Geber GE 203 joystick controller onboard the host aircraft to steer the missile. The joystick was popularly called the Knüppel (club), a device that would eventually become a common feature on Luftwaffe guided weapons.

The SC 500 high-explosive fragmentation bomb had been selected as the basis for the missile. This contained a 300kg (650lb) high-explosive charge detonated by an impact fuze. Unlike the Fritz-X, the Hs 293 warhead had very limited armor penetration power, and the missile was intended for the attack of lightly armored warships such as destroyers, or merchantmen. Serial production of the Hs 293A-0 began in 1942, totaling 1,280 missiles, before switching in November 1942 to the improved Hs 293A-1, which had upgraded electronics.

The Hs 293A-1 became the standard production version, with 5,923 built from November 1942 to the end of 1943 and more than 3,500 in 1944. Most of the production was undertaken at Henschel's Werk III in Niederschöneweide in the southwest suburbs of Berlin, with about 11,400 missiles of all versions completed by the end of July 1944 when production ceased. Many of the subvariants of the Hs 293 described later were originally built as Hs 293A-0 or A-1, but then converted. A total of about 300 bombers were modified with Kehl transmitters and other features to carry the Hs 293 and Fritz-X, including 140 Do 217E-5s, K-2s, and M-11s, as well as 107 He 177s, 40 Fw 200s, and small numbers of Ju 290A-7s and He 111s.

B **THE Hs 293 FAMILY**

1. The Hs 293A-1 was the most widely produced version of the series.

2. The Hs 293C was fitted with the "torpedo" nose of the larger Hs 294 series. The Hs 293C-3 was the wire-guided version, evident from the wire bobbins fitted at the wingtips.

3. The Hs 293D was fitted with a camera in the nose for terminal guidance and a Yagi antenna in the tail to improve data transmission and reception.

4. The Hs 294 was the "torpedo" version of the series, with a special conical warhead designed to travel underwater to the target and strike the ship under the waterline. Due to its increased weight, it had two rocket boosters.

5. The Hs 295 was an enlarged version of the family with a heavy, armor-piercing warhead.

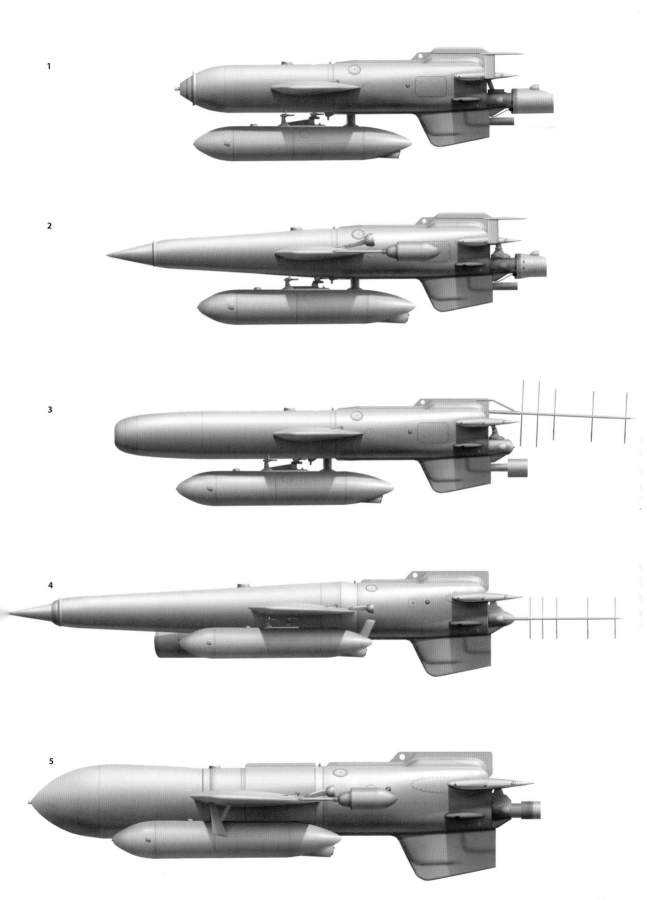

1

2

3

4

5

Antiship Missiles Comparative Technical Data

	Fritz-X	Hs 293A1
Length (m/ft)	3.26/10.7	3.88/12.7
Diameter (m/ft)	0.56/1.8	0.47/1.5
Wingspan (m/ft)	1.35/4.4	3.1/10.2
Warhead (kg/lb)	1,400/3,085	510/1,120
Engine	None	HWK-109-507B rocket
Weight (kg/lb)	1,565/3,450	785/1,730

Into Combat

In early 1943, the two missile test squadrons were enlarged and deployed with the specialist antishipping wing, Kampfgeschwader.100 "Wiking". The EK.15, specializing in the Hs 293, became II./KG.100, while EK.21, specializing in the Fritz-X, became III./KG.100. In July 1943, the two groups were deployed to southern France for missile operations against Allied shipping in the Mediterranean. The III./KG.100 began attacks using the Fritz-X on July 21 against Augusta harbor and on August 1 against the port of Palermo, both in Sicily, but without success. The attacks against Allied ports in the Mediterranean continued through early August and September, but no Allied ships were sunk.

The first use of the new Hs 293 missiles occurred on August 25, 1943, when a dozen Do 217E-5s of II/.KG.100 attacked a small Royal Navy force in the Bay of Biscay, damaging two small warships with near misses. On August 27, another attack was staged in the same area against the Royal Navy's 40th Escort Group. The eighteen Do 217s were each carrying a single Hs 293. The sloop HMS *Egret* was attacked by seven missiles. It dodged five and shot down one, before being struck and sunk by the seventh. Five bombers attacked the destroyer HMCS *Athabaskan*, managing to hit it once. The missile's warhead penetrated the thin armor on both sides and passed through the ship before detonating beyond.

The most spectacular victory with the new guided weapons occurred on September 9, 1943. When the Italian government decided to switch sides and abandon Germany, Allied negotiators insisted that the Italian Navy remove its three modern battleships from La Spezia near Genoa to prevent them from falling into the hands of the Kriegsmarine. The Luftwaffe learned of the plans. When the sortie took place on September 9, six Do 217K-2 aircraft, each armed with a single Fritz-X, set out to sink the battleships. The attack took place to the west of the Strait of Bonifacio, between Corsica and Sardinia. The battleship RM *Roma* was struck by a Fritz-X that damaged the engine room without detonating until it careened outside. The coup-de-grace was a second high-angle strike into the forward magazine near the B turret, leading to a catastrophic explosion that tore off the turret, detonated the ammunition, and destroyed the ship, killing most of its crew. The battleship *Italia* was struck by a Fritz-X that also went through the hull before exploding beyond, but the ship survived.

Over the course of the next three weeks, the two "Wiking" groups took part in a series of air attacks against the Allied amphibious landings at Salerno in southwestern Italy. The cruiser USS *Savannah* was hit by a Fritz-X on its third forward turret, penetrating into the magazine before exploding. The detonation and subsequent fire killed 197 men, including the turret crew. Days later, two Fritz-X missiles penetrated the rear decks of the battleship

HMS *Warspite*, severely damaging the powerplants but missing any of the highly combustible magazines. Several small ships were sunk and damaged in the attacks.

The Salerno campaign was the swansong of the Fritz-X. The Fritz-X was not well suited to attacking most common naval targets such as destroyers or unarmored merchantmen, because the weapon usually passed through the hull before exploding. In addition, the Fritz-X had to be released near the ship, increasing its chances of being hit by naval antiaircraft fire. The III./KG100 suffered heavy losses in the campaign and had only five remaining aircraft. Subsequent attacks were conducted largely with Hs 293s, which could be launched from greater stand-off range.

Missile attacks in late September included missions against Ajaccio harbor on Corsica. One of the Do 217 bombers lost during this short campaign was a special version fitted with electronic intelligence devices to determine whether the Allies were attempting to jam the Kehl-Straßburg command link. No such jamming signals were detected. In the absence of any such evidence, Henschel put plans for a wire-guided version of the Hs 293 on the back shelf. One of the deadliest attacks was the sinking of the troop ship HMT *Rohna* in the Mediterranean on November 26, 1943, by an Hs 293 guided missile. More than 1,100 people were killed, mostly US soldiers.

The Allied responses to the guided weapons attacks were both tactical and technical. The bombers' bases were bombed, destroying both aircraft and missile stores. Fighter patrols over the Salerno beachhead were increased. In September 1943, the US Navy commissioned two new destroyer escorts,

A remarkable photo taken moments after the cruiser USS *Savannah* was struck by a Fritz-X on one of its forward turrets on September 11, 1943, off Salerno. The bomb killed most of the turret crew, but damage control parties managed to save the ship.

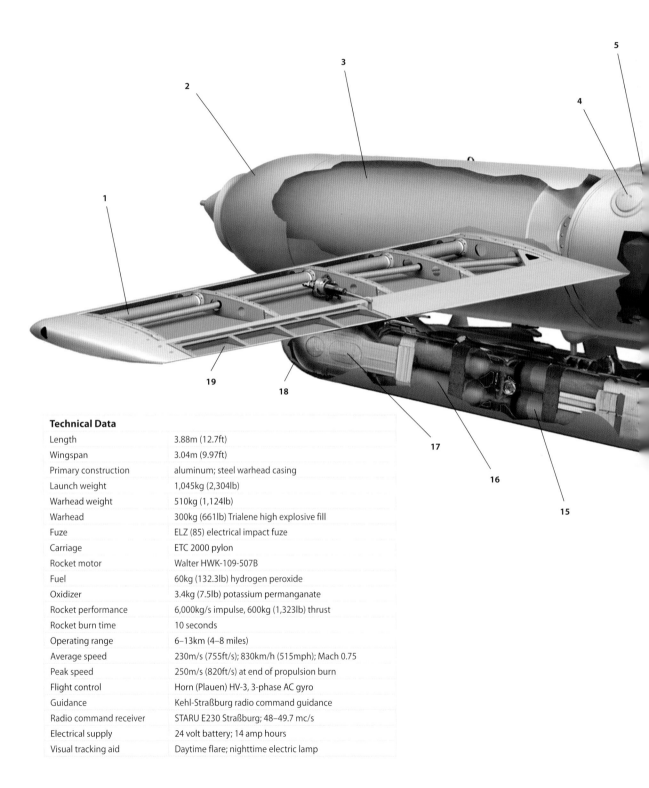

Technical Data

Length	3.88m (12.7ft)
Wingspan	3.04m (9.97ft)
Primary construction	aluminum; steel warhead casing
Launch weight	1,045kg (2,304lb)
Warhead weight	510kg (1,124lb)
Warhead	300kg (661lb) Trialene high explosive fill
Fuze	ELZ (85) electrical impact fuze
Carriage	ETC 2000 pylon
Rocket motor	Walter HWK-109-507B
Fuel	60kg (132.3lb) hydrogen peroxide
Oxidizer	3.4kg (7.5lb) potassium permanganate
Rocket performance	6,000kg/s impulse, 600kg (1,323lb) thrust
Rocket burn time	10 seconds
Operating range	6–13km (4–8 miles)
Average speed	230m/s (755ft/s); 830km/h (515mph); Mach 0.75
Peak speed	250m/s (820ft/s) at end of propulsion burn
Flight control	Horn (Plauen) HV-3, 3-phase AC gyro
Guidance	Kehl-Straßburg radio command guidance
Radio command receiver	STARU E230 Straßburg; 48–49.7 mc/s
Electrical supply	24 volt battery; 14 amp hours
Visual tracking aid	Daytime flare; nighttime electric lamp

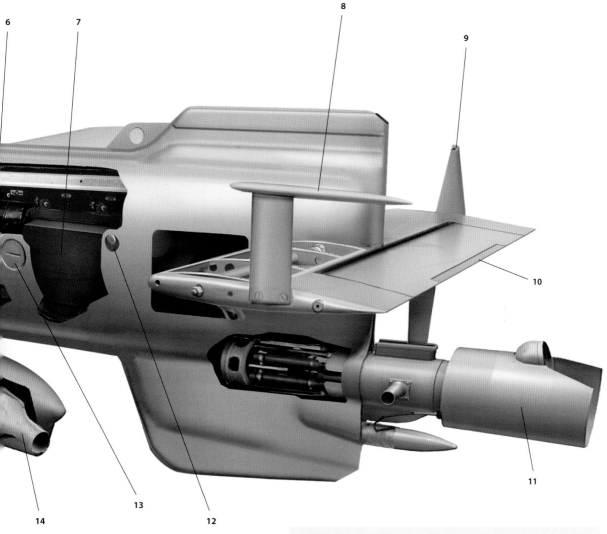

6 **7** **8** **9**

10

11

13

14 **12**

KEY

1. Main wing spar
2. Warhead casing
3. High explosive warhead
4. Hot air intake (from parent aircraft)
5. Guidance filter unit
6. DC generator
7. STARU E230 Straßburg radio command receiver
8. Main antenna support
9. Antenna support

10. Elevator
11. Tail tracer flare assembly
12. Guidance tuning window
13. Radio destruction device
14. Rocket combustion chamber
15. Air pressure tank
16. Z-Stoff tank
17. T-Stoff tank
18. Walter HWK-109-507B rocket motor
19. Aileron

the USS *Herbert C. Jones* and USS *Frederick C. Davis*, that were fitted with experimental jammers. These first saw use in early November, escorting convoys in the Mediterranean, but they were not effective as the precise operating frequencies of the Kehl-Straßburg were not yet understood. At this stage in the missile war, preemptive strikes against the German airbases proved more effective than the radio jammers.

In late November, the two missile groups of KG.100 "Wiking" were temporarily withdrawn for re-equipping and rebuilding after three costly months of fighting in the Mediterranean. The II./KG.100 had started the campaign with 43 Do 217E-5 bombers, losing six in combat and 27 from attrition and accidents. The II./KG.100 had started with 35 Do 217K-2s, losing 17 in combat and 22 to accidents and attrition. The 7./ and 9./KG.100 were sent to Eggebek near the German North Sea coast as part of Operation *Carmen*, a still-born scheme to hit the Royal Navy's Home Fleet in Scapa Flow. Other squadrons were earmarked for raids against Soviet industrial targets that also failed to materialize.

Besides the veteran KG.100 "Wiking", II./KG.40 had converted to the new He 177A-3 Greif heavy bomber in September 1943, fitted with the Hs 293 missile. They saw their combat debut in late November 1943 off the western coast of Britain, attacking Allied convoys. Combat losses and teething problems with the temperamental new aircraft led to the loss of about half the force in November–December 1943, totaling 25 aircraft, for very modest successes.

The next spasm of missile attacks in January 1944 were centered on the Mediterranean, targeting the Allied amphibious landing at Anzio. The II./KG.100 had about 45 Do 217E-5 bombers at the outset of the campaign. The new II./KG.40 also took part in the attacks. The first victim of the Hs 293 in this campaign was the destroyer HMS *Janus* on January 23. The threat of missile attacks led to more vigorous Allied air patrols that paid off on the night of January 26, when US fighters from Corsica intercepted the inbound He 177 bombers of II./KG.40, claiming six bombers. This group was subsequently withdrawn from the hotly contested Anzio air battles and sent to prey on easier convoy targets. The missile groups lost about half their aircraft in the air battles over Anzio in January 1944. Overall, KG.100 was far less successful at Anzio than at Salerno. The US Navy attributed this to the presence of three new destroyers fitted with jammers, though it can certainly be argued that vigorous Allied fighter cover was more instrumental in blunting the missile attacks.

By late autumn 1943, the Allies were finally beginning to piece together the operating frequencies of the Kehl-Straßburg system. The Allies found an intact Hs 293 on Anzio beach with an undamaged Straßburg receiver. Besides determining the operating bands, this permitted the Allies to determine how the signals actually guided the missile. This meant that in the future, the Allies could not only jam the frequencies, but also spoof the guidance channel with false commands.

Through April 1944, the Luftwaffe conducted 487 missile sorties, of which 313 reached the target area. These claimed to have scored 66 direct hits and 40 near misses with the Hs 293 and Fritz-X, sinking 30 ships and damaging 51. According to Martin Bollinger's study of the German antiship missiles, actual hits were 29 and near misses 14; casualties were 19 ships sunk and 18 damaged. It should be noted that the source of some Allied losses

KG.100 attempted to counter-attack the Allied amphibious invasion on the Normandy coast in June–July 1944, but Allied fighter coverage quickly decimated the bomber formations. In early August 1944, the attacks shifted to the bridges near Avranches in support of Operation *Lüttich*, the German attempt to stem the Operation *Cobra* breakout. One of these night attacks is depicted in this illustration. (Author)

were not certain, since missile attacks often occurred along with Luftwaffe bomb and torpedo attacks. About 700 missiles and guided bombs were launched in this period, of which about a third failed to respond to guidance signals. The average success rate of the missiles that were launched was only 7 percent, and even if the high rate of guidance failures is discounted, the average hit/near-miss success rate was only 11 percent.

The Normandy Invasion

By early 1944, more He 177A-5 bombers were deployed for missile operations, as well as some Fw 200s with III./KG.100 and Ju 290s with Fernaufklärungs-Gruppe.5. At the time of the Allied landings in Normandy in June 1944, the Luftwaffe had four groups deployed with the guided weapons, including elements of the newly converted I./KG.100. In spite of the new aircraft, the actual capabilities of the bomber squadrons had been reduced by combat attrition and the lack of experienced crews. Nevertheless, Luftwaffe chief Hermann Göring still called these four groups "the spearhead of the anti-invasion force." In practice, their performance against Operation *Overlord* would be very disappointing.

The secrets of the German missiles continued to unravel as the Allies recovered another intact Hs 293 on Corsica, and decrypted a technical briefing on missile guidance transmitted from the Japanese embassy in Berlin. Besides the availability of more refined electronic countermeasures, the real Allied advantage was in air power, both the availability of ample fighter protection over the beachhead and a robust bombing campaign against all known German missile-bomber bases.

The first missile attack took place on the night of June 6–7, with He 177s from KG.40 arriving near the Normandy landing sites shortly before dawn. They had been spotted by Allied radar; USAAF and RAF night fighters embarked on "the slaughter of the Heinkels." At least six were lost and only one missile came near any Allied ships. The early morning attacks on June 8 saw the loss of four more bombers, but the USS *Meredith* was probably damaged by a Hs 293 before being sunk later in the morning in a conventional bombing attack. The converted destroyer/command ship HMS *Lawford* also suffered a missile hit. Over the next week, the missile aircraft scored a few hits, but the bomber losses were staggeringly high. The II./KG.40 lost 23 bombers in combat and seven to accidents and attrition; III./KG.100 lost 11 in combat and 12 to accidents and attrition. KG.40 lost about 80 percent of its aircraft in June, mostly in the first week after the landings. The Allied defenses over the *Overlord* beaches forced the Luftwaffe to shift the focus of the attack away from Normandy and back to the Bay of Biscay. The frigate HMCS *Matane* was heavily damaged off Brest. The last victim of the German missiles was LST-282, hit off the Provence coast during the Operation *Dragoon* landings in southern France on August 15.

Hitler's fury over the Allies' breakout from Normandy in the wake of Operation *Cobra* in late July 1944 led to one of the few major uses of the Hs 293 in attacks against land targets. Gen George Patton's Third US Army had spilled westward into Brittany over the Pontaubault bridges, and Hitler insisted that the bridges be destroyed. The attacks began on the night of August 2 and lasted through the night of August 6–7; none of the bridges were hit and the bomber squadrons lost five more aircraft.

Most of the missile groups were withdrawn from France and gradually disbanded through the autumn. There were a handful of missile missions in 1945, including last-ditch attacks by KG.200 on the Oder River bridges on April 12, 1945, using the Hs 293. The Hs 293 was not powerful enough for some of the bridges, leading to the use of Mistel assault drones. These consisted of surplus Ju 88G night fighters radio-controlled by accompanying Fw 190 fighters. Although the Mistel can be rightly viewed as a guided missile, this program is not covered in detail in this book for space reasons as well as the fact that it has been covered in detail in another Osprey book.[2]

Overall, the performance of the antiship missiles was disappointing except for the rare successes such as the sinking of the *Roma*. The Luftwaffe flew 903 missile sorties in 1943–44, losing 112 bombers prior to launch and 21 on the return flight for an average loss rate of 15 percent per sortie. The rate of success changed through time, partly as a result of improved Allied countermeasures. Through January 1944, successful launch claims averaged 20–30 percent of missiles that did not malfunction. By February 1944, when a greater number of jammers were available to the fleet off Anzio, the success claims tumbled to about 10 percent, eventually falling to less than 5 percent in the summer of 1944. The more experienced KG.100 "Wiking" generally had higher success rates than the novice KG.40 which had only two hits off Normandy out of 130 sorties. No single factor accounts for the declining performance of the missiles. Rather, it was a combination of deteriorating Luftwaffe crew experience, more vigorous Allied fighter defense, and the greater presence of sophisticated shipboard jammers.

---

2   Robert Forsyth, *Luftwaffe Mistel Composite Bomber Units*, Osprey Combat Aircraft 112, Oxford (2015).

German Missile Strikes on Allied Ships 1943–44

Date	Ship	Type	Weapon
25 Aug 43	HMS *Landguard**	cutter	Hs 293
25 Aug 43	HMS *Bideford**	sloop	Hs 293
27 Aug 43	HMCS *Athabaskan**	destroyer	Hs 293
27 Aug 43	HMS *Egret***	sloop	Hs 293
9 Sep 43	RM *Roma***	battleship	Fritz-X
9 Sep 43	RM *Italia**	battleship	Fritz-X
11 Sep 43	USS *Savannah**	cruiser	Fritz-X
13 Sep 43	HMHS *Newfoundland***	hospital ship	Hs 293
13 Sep 43	HMS *Uganda**	cruiser	Fritz-X
13 Sep 43	HMS *Loyal**	destroyer	Fritz-X
15 Sep 43	SS *Bushrod Washington***	cargo ship	Hs 293
15 Sep 43	SS *James W. Marshall***	cargo ship	Hs 293
17 Sep 43	HMNLS *Flores**	gunboat	Fritz-X
18 Sep 43	HMS *Warspite**	battleship	Fritz-X
30 Sep 43	HMS *LST-79***	landing ship	Hs 293
4 Oct 43	SS *Samite**	cargo ship	Hs 293
4 Oct 43	SS *Hiram S. Maxim**	cargo ship	Hs 293
4 Oct 43	SS *Selvik**	cargo ship	Hs 293
11 Nov 43	MV *Birchbark***	cargo ship	Hs 293
11 Nov 43	MV *Indian Prince***	cargo ship	Hs 293
11 Nov 43	HMS *Rockwood ***	destroyer	Hs 293
11 Nov 43	HMS *BYMS-2072**	minesweeper	Hs 293
13 Nov 43	HMS *Dulverton***	destroyer	Hs 293
21 Nov 43	MV *Marsa***	cargo ship	Hs 293
21 Nov 43	SS *Delius**	cargo ship	Hs 293
25 Nov 43	HMT *Rohna***	cargo ship	Hs 293
23 Jan 44	HMS *Janus***	destroyer	Hs 293
24 Jan 44	HMHS *St David***	hospital ship	Hs 293
24 Jan 44	USS *Prevail**	minesweeper	Hs 293
24 Jan 44	USS *Mayo**	destroyer	Hs 293
26 Jan 44	SS *John Banvard***	cargo ship	Hs 293
26 Jan 44	SS *Hilary A. Herbert**	cargo ship	Hs 293
28 Jan 44	HMS *Jervis**	destroyer	Hs 293
29 Jan 44	SS *Samuel Huntington***	cargo ship	Hs 293
29 Jan 44	HMS *Spartan***	cruiser	Hs 293
15 Feb 44	SS *Elihu Yale*/LCT-35**	landing craft	Hs 293
25 Feb 44	HMS *Inglefield***	destroyer	Hs 293
10 Jun 44	SS *Fort McPherson**	cargo ship	Hs 293
7 Jun 44	USS *Meredith**	destroyer	Hs 293
8 Jun 44	HMS *Lawford***	frigate	Hs 293
10 Jun 44	SS *Charles Morgan***	cargo ship	Hs 293
13 Jun 44	HMS *Boadicea***	destroyer	Hs 293
20 Jul 44	HMCS *Matane**	frigate	Hs 293
15 Aug 44	USS LST-282**	landing ship	Hs 293

*Damaged

**Sunk, scuttled, written off

Improved Antiship Missiles

Henschel continued to develop improved versions of its antiship missiles, though none reached serial production. Due to concern over possible jamming of the Kehl-Straßburg radio link, about 200 missiles were built in the Hs 293B configuration with a wire guidance system. This consisted of an S207 Dortmund transmitter on the aircraft and an E237 Duisberg receiver on the missile. Six Do 217 and He 177 bombers were converted for the developmental trials by EK.36, but so far as is known, no German bombers with these features were deployed into combat because there was widespread skepticism about Allied jammers.

Another approach to guidance was the use of a television camera in the nose of the missile to assist the bombardier in steering the missile during the terminal approach to the target. Tests of the Hs 293D began in 1942 but were plagued by the poor reliability of the primitive K-11 camera. Improvements to the cameras led to a new series of tests in October 1944, but by this time, interest in antiship missiles was waning due to a lack of bombers.

Later versions of the Hs 293 family included the Hs 293E with improved radio controls, and the HS 293F with a tailless delta wing configuration and solid-rocket booster. A submarine-launched version was also under development. None of these reached serial production.

In parallel to the Hs 293, Henschel worked on the Hs 294, called a guided torpedo bomb. This was not a torpedo in the usual sense. It used a special conical warhead, configured to permit the missile to plunge into the water before hitting the ship, with the warhead breaking away from the remainder of the missile and striking the target underwater. The Luftwaffe hoped that such a warhead could enter the water 500–700 meters (1,650–2,300ft) away from the ship, but in practice, it could only travel underwater about 45m (145ft). The Hs 294 had two rocket boosters instead of the one on the standard Hs 293. Few of the Hs 294A-1 missiles were manufactured. The underwater-attack mode led in two other directions. The Hs 293C variants used the standard Hs 293 configuration with a single rocket booster, but fitted with a conical "torpedo" warhead; several subvariants were developed. The Hs 295, tested in the summer of 1943, was an enlarged Hs 293 using the twin booster configuration of the Hs 294 as well as an armor-piercing warhead. The Hs 296 was a further elaboration of the Hs 295 program, using a BK 1400

armor-piercing warhead and first tested in April 1944 for possible use from the Ar 234 jet bomber as an alternative to the Fritz-X.

The Fritz-X went through continual evolutionary design. The X-2 used improved guidance. The X-3 introduced a new airfoil for spin stabilization during descent. This configuration was adapted to 2,500kg bombs, the X-5 using an armor-piercing bomb and the X-6 a high-explosive/fragmentation bomb. A contract to build 100 of both types was quickly canceled. A smaller version of the Fritz X-5, the Peter X, was also considered.

FLAK MISSILES

The development of Flak missiles (Flakraketen) began in the early years of the war but did not receive significant resources until late 1943. These efforts were first undertaken by the Luftwaffe's Technisches Amt (Technical Development Branch), better known as the C-Amt. In the late 1930s, the C-Amt created a new branch headed by former Heinkel engineer Rudolf Brée to manage the development of Luftwaffe guided weapons including air-to-air, surface-to-air and torpedoes. There was an affiliated office, headed by Major Dr Friederich Halder, at the Army's Peenemünde missile establishment on the Baltic to conduct weapons trials.

By 1941, the Luftwaffe realized that conventional Flak artillery was approaching the limit of its capabilities to deal with fast, high-altitude bombers operating at night. In February 1941, Maj Halder proposed the development of guided Flak missiles as a technical solution. At the time, there was little support for the heavy investment that would have been required for a Flak missile program.

This Mistel consisted of a war-weary Ju 88G night fighter converted into a radio-guided flying bomb with a Fw 190 fighter attached above to guide it to the target area. The fighter roughly aimed the drone at the target, separated from the combination, and then guided the drone until impact. This illustration shows one of the Mistel operations by KG.200 against the Oder River bridges in early April 1945. (Author)

OPPOSITE
The Blohm & Voss Bv 246 Hagelhorn was offered to the Luftwaffe as a cheaper alternative to the FZG-76/V-1 for aerial bombardment of large targets such as cities. Other versions were proposed, including antiship and antiradar types. About 400 were manufactured, but they were set aside as target drones for testing the new Flak missiles. (Author)

The Ardelt F25 Feuerlile (Fire-Lily) was one of a number of experimental missiles developed to study transonic speeds. It was first tested in April 1943, but was never considered for operational use.

On September 11, 1941, Hitler ordered the cessation of all long-term research programs in favor of concentrating on existing development programs. This further reduced the chances for a near-term Flak missile program since no formal effort had yet been approved. It also had long-term consequences since it reduced the funding for radar and proximity fuze development that were essential for any long-range Flak missile effort. The C-Amt further dampened the prospects for Flak missiles after releasing a technical memo that argued that air-to-air guided missiles for fighter aircraft would be a more economical alternative to Flak missiles. This inspired Luftwaffe chief Hermann Göring to issue an edict canceling development of Flak missiles.

In spite of the lack of enthusiasm of many senior Luftwaffe officials, the inspector of Flak artillery, Gen Lt Kurt Steudemann, continued to press for development of Flak missiles, fully realizing it would be a time-consuming process. Gen Walter von Axthelm, appointed to replace Steudemann in January 1942, was also a very vigorous supporter of Flak missiles. He began conducting presentations to Göring and other senior Luftwaffe officials, trying to build up political support. At Axthelm's instigation, Halder wrote up another proposal in in April 1942. He followed this with a May 1942 report that berated the Amtsgruppe für Flakentwicklung (Flak Development Office). Flak development had been under Army control until transferred to the Luftwaffe in 1940, and Halder characterized it as "a collection of out-of-touch Army traditionalists who failed to see the potential of radical new technologies such as the missile."

It took Axthelm several months to convince Göring, who finally approved the start of the "R-Programm" (R: Rakete – Rocket) on September 1, 1942, followed by Hitler's belated blessing. In June 1942, Gen der Flak Otto-Wilhem von Renz took over the Amtsgruppe für Flakentwicklung, and the new missile program was taken from the unsympathetic hands of C-Amt and transferred to Abteilung E (Flak) under Renz's supervision, more formally designated OKL-TLR Flak E. Halder was put in charge of the Flak missile office at Peenemünde, known as Flak E5.

A Flak rocket narrowly misses a B-17G during a mission over the Rhine river on January 13, 1945.

In practice, the Luftwaffe had enormous difficulty in recruiting a sufficient number of talented engineers for such an ambitious program. Halder's small Flakkommando at Peenemünde was heavily dependent on the Army's missile program under Gen Dornberger that was managing the A-4 (V-2) ballistic missile program. To some extent, the Luftwaffe's impoverished and underfunded R-Programm was able to piggy-back on the much more lavishly funded Army ballistic missile program.

Unguided Flak Rockets

In contrast to the FZG-76 (V-1) cruise missile and A-4 (V-2) ballistic missile, Hitler showed no particular enthusiasm for the Flak missile program and it remained on shaky political grounds in Berlin with inadequate funding and low priority for personnel. The program was further disrupted in August 1943 when Hitler insisted that the Luftwaffe deploy the army's 21cm Nebelwerfer multiple-rocket artillery launcher for air defense of the Ruhr industrial region in the face of the RAF's bomber campaign. This weapon was completely unsuitable for this role since the rockets did not have the range or speed to engage aircraft. The Luftwaffe Flak branch was aghast at this idea, but also realized that if they did not respond to Hitler's demand, the whole Flak missile program would be put into jeopardy. As a result, the Luftwaffe began development of the 21cm Maikäfer (May bug) Flakrakete 42 rocket launcher at Peenemünde. Although it went into limited production, it proved entirely worthless against Allied bombers.

The R-Programm adopted a three-stage approach to the development of Flak missiles. The first stage, and the simplest, was a revival of the barrage rocket approach, using multiple unguided rockets optimized for antiaircraft use. This was intended for low-altitude defense of air bases and other objectives since there was the realization it would not have the range to deal with the high-altitude bomber threat.

This first emerged as the 7.3cm Föhn Flakwerfer 44. Test launches began on October 11, 1943, and the system entered serial production in October 1944. There were plans to ramp up production to 200 launchers monthly by the end of 1944. In the event, only 85 Flakwerfer 44 Föhngeräte were built by the time production ended in February 1945.

One of the few unguided Flak rockets to enter service was the Flakwerfer 44 Föhngeräte, which could fire 35 unguided 73mm rockets. This example served with the I./Flak Lehr und Erprobungs Abt.900 in Erpel, defending the Ludendorff bridge at Remagen. It is seen here after its capture by the US Army with an M16 antiaircraft halftrack of the 634th AAA Battalion (Automatic Weapon) behind it.

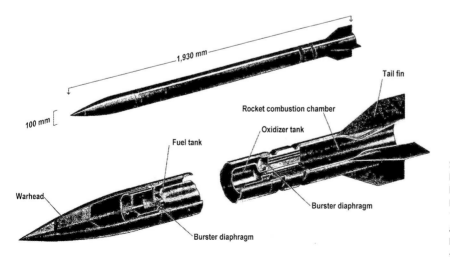

1,930 mm

100 mm

Warhead

Fuel tank

Rocket combustion chamber

Oxidizer tank

Tail fin

Burster diaphragm

Burster diaphragm

Since the 73mm Föhn did not have enough range to reach bombers, EMW in Karlshagen/Peenemünde developed the 100mm Taifun (Typhoon) antiaircraft rocket using a liquid-fuel rocket engine, as shown in this cut-away.

Many of the test-bed launchers were deployed at air bases, with a few being mounted on rail-cars for mobility. The Föhngeräte was supposed to be followed by more powerful liquid-fueled unguided rockets. The most refined of these was the 10cm Taifun liquid-fuel rocket developed at the Elektromechanische Werke in Karlshagen, the manufacturing arm of the Peenemünde establishment. This had a range of about 15km (9 miles) and a maximum altitude of 12,000m (39,350ft). A 30-cell launcher was mounted on a cruciform platform derived from the 88mm Flak gun. Production of 2 million rockets and 400 launch batteries was supposed to begin in January 1945, though only test batches were ever completed. A third program, the Fliegerfaust, was a shoulder-launched weapon for troop defense against low-flying Allied aircraft. It saw limited service in the final weeks of the war.

Flak Missiles

The second stage of the R-Programm was the development of a subsonic missile using electro-optical tracking and radio command guidance. This was essentially similar in concept to the Kehl-Straßburg guidance used with Luftwaffe antiship missiles. However, due to the greater range of operation and the smaller size of the target, a more elaborate optical tracking system and more powerful transmitters were needed. The German electronics firm Telefunken was assigned the task of developing the common fire-control systems for these missiles and supervising various companies and government institutes designing key subcomponents such as proximity fuzes.

The third phase of the program was the development of a supersonic missile using radar fire controls. Due to the speed of the missile, the presumption was that electro-optical command guidance would be inadequate and that some more sophisticated form of guidance would be needed. A variety of guidance methods were explored, including infra-red homing, radar beam-riding, and semi-active radar homing. This represented the single most challenging technology of the war.

By early 1943, there were nearly four dozen assorted missile and guidance programs. Many of the smaller programs were weeded out during the course of 1943. Only four of the missile programs received extensive government funding. This consisted of three subsonic missiles – the Henschel Schmetterling, Rheinmetall-Borsig Rheintochter, and Messerschmitt Enzian – and the supersonic Wasserfall missile. These four programs are the focus of this book since they came closest to service deployment.

The plans were to deploy a multi-layered Flak missile network to defend all of Germany against strategic bomber attack. The unguided rockets would form an initial, low-altitude barrier. This would be followed by 1,300 batteries of one of the medium-range, subsonic missiles such as Schmetterling. Finally, there would be a further 870 batteries of long-range, supersonic Wasserfall missiles. This would require a stupendous financial and industrial investment in launch sites, missiles, and radars. Early

---

D **RHEINTOCHTER III LAUNCH SITE**

The Rheintochter III was designed to be launched from fixed, traversable launchers. The launcher was mounted in a shallow concrete "kettle" position. To facilitate rapid reloading, a narrow-gauge railway track lead to the rear of the launcher, enabling a small reloading dolly to be moved up to the launcher.

estimates put monthly production of the Wasserfall at 10,000 missiles. The Flak missile batteries would require a force of 110,000 personnel.

The precision guidance requirements for the Flak missiles were infinitely more demanding than for the existing V-1 and V-2 surface-to-surface missiles. The vengeance weapons used inertial guidance platforms based on gyroscopes which were accurate to within several kilometers against a static target. The Flak missiles needed guidance system accurate to within several meters of a moving target, a thousand-fold difference. The German missile guidance systems of 1944–45 could not possibly hope for a direct impact with the target, so some form of proximity fuze was needed to detonate the missile warhead when it passed nearest to the target. Germany had poor results in this field compared to the United States and Britain. German organizations pursued a variety of technologies, including optical, acoustic, and radio frequency (RF) fuzes, though none was fully mature when the war ended.

Rheinmetall-Borsig Rheintochter

The Rheintochter (Rhine's Daughter) was developed by Rheinmetall-Borsig AG, headquartered in Berlin-Marienfelde, under the direction of Dr Ing Heinrich Klein. Rheinmetall had already submitted a proposal to the RLM in December 1941 but there was no Luftwaffe interest. After Göring's September 1942 decree starting the R-Programm, Rheinmetall resubmitted the proposal and was awarded a development contract on December 7, 1942.

Development began in early 1943 using a half-scale model, launched from a test range at Leba on the Baltic coast. This pioneered a two-stage, solid-fuel missile configuration. The booster stage operated for 1.2 seconds, followed by the sustainer motor. Up to this point, solid-fuel rocket engines had only 10–12-second burn times, but this missile would require burn times on the sustainer motor of up to 60 seconds, a substantial technological challenge. A full-scale version of the Rheintochter I was completed in May 1943. The initial launch on November 14, 1943, was an unguided ballistic test. Rheintochter II was a refined version of the Rheintochter I, but did not progress beyond the design stage.

The early test launches of the Rheintochter I were plagued by stability problems, leading to several redesigns of the fins. By the end of 1943, six test launches had taken place, followed by two more in January 1944, mainly from the Leba test site on the Baltic. Test launches of the Rheintochter I continued through early 1945, mainly to demonstrate subsystems and guidance. Of the 82 test missiles that were launched, 22 used the radio command guidance system while the others were simple ballistic tests. A total of about 100 Rheintochter I missiles were built.

In September 1943, Albert Speer's RBM armaments ministry informed Rheinmetall that it would not support the program due to the excessive consumption

Fighter production had priority over Flak missiles, so many designs were obliged to use wood instead of aluminum where possible. This is evident on the wings of this Rheintochter I on display at the Udvar-Hazy Center of the National Air and Space Museum. (Author)

of scarce propellant required for each missile. In addition, the Luftwaffe was disappointed by the relatively poor performance of the missile in terms of range and altitude. As a result, the missile had to be redesigned to use a liquid-fuel rocket engine as the Rheintochter III. Gen.von Renz attributed the problems to the lack of experienced aeronautical engineers with the firm, which had previously been engaged mainly in ammunition and gun design. He attempted to get several experienced aeronautical engineers from the C-Amt but was rebuffed, which caused serious delays in the program.

The Rheintochter I used a modified carriage from an 88mm Flak 41 for its test launcher.

The Rheintochter III began to take shape in early 1944. It used two smaller solid-fuel boosters and the VFK 613 liquid rocket sustainer, patterned after the engine used in the Wasserfall missile. Test launches were delayed until late October 1944. Through January 1945, only six launches were conducted, five with a stop-gap solid-fuel sustainer and only one with the new VFK 613 liquid-fuel engine. None of these tests were guided.

In late 1944, there was some effort to narrow down the number of subsonic Flak missiles under development in order to concentrate on at least one design for serial production. The Kommission für Fernschießen (Long-Range Weapons Commission), headed by Prof Waldemar Petersen of the AEG electronics firm, recommended that work on the Rheintochter and Enzian be halted since they had not progressed sufficiently and would probably be prohibitively expensive. In the event, the Rheintochter emerged as one of the least successful of the major Flak missile programs, in no small measure due to the need to redesign the missile in the midst of development. The program was finally canceled on February 6, 1945.

Henschel Hs 117 Schmetterling

Henschel Flugzeuge Werke AG near Berlin was one of the few German firms with actual experience in missile design due to Prof Herbert Wagner's previous work on the Hs 293 antiship missile. Wagner had studied the idea of adapting the Kehl-Straßburg guidance technology to a Flak missile as early as 1941. He proposed the Hs 297 to the RLM in 1941, but it was rejected as unnecessary. Following the start of the R-Programm, Henschel was given a contract for a small Flak missile under the Luftwaffe codename Schmetterling (Butterfly).

By this time, Wagner was Germany's most experienced tactical missile designer and appreciated the many factors necessary to build an expendable missile. Rather than aim at the highest level of technological excellence, he sought to develop a missile that was small and cheap while still offering good performance. He chose a design that could be developed as quickly as possible. The Hs 297 Schmetterling was considerably smaller than its competitors since Wagner selected a small warhead, accepting that on average it would take three missiles to shoot down a typical target, namely a B-17 flying at 335 mph(540km/h) at a height of 6,000–8,000m (19,500–26,250ft) and a range of 15km (9 miles).

The Henschel Hs 117 V-1 Schmetterling used a traversable launcher during the test phase.

The Schmetterling missile resembled a small aircraft. The missile was initially propelled into the air by a pair of Schmidding 109-533 di-glycol solid-rocket boosters. The BMW 109-558 liquid-fuel rocket was used as the sustainer engine, but its inadequate thrust led to consideration of the Walter HWK109-729 as an alternative.

The Schmetterling nose had an unusual asymmetric shape, which was repeated on other Henschel missiles. On one side was a protruding cone that contained a proximity fuze, while to the side and slightly behind was a small airscrew that was used to drive a generator that provided electrical power for the missile's flight control system. The use of a generator avoided the need for maintaining a battery when the missile was stored. The nose section also contained the high explosive-blast warhead. The flight control system was very similar to the Hs 293, using gyroscopes for roll control. One major difference was the use of the lighter E232 Colmar radio receiver in the missile instead of the heavier Straßburg receiver.

Flight tests of the Schmetterling began at Peenemünde-West in May 1944. In July 1944, the Luftwaffe changed the designation of the missile from

E **Hs 117 SCHMETTERLING LAUNCH SITE**

Had the Schmetterling gone into series production, it was expected to use a new type of Startgestell (launcher). This was a zero-length launcher depending on the impulse of the booster rockets to get the missile airborne. It consisted of a traversable launch array on a pedestal mount, with a cruciform base. The launcher would be reloaded from a Zubringewagen, a type of cart to carry the missile from the preparation sheds to the launcher. Guidance of the missile was undertaken from a two-man Burgund Flakfolgegerät, usually positioned about 40m away from the launchers and seen here to the left.

The Henschel Hs 117 Schmetterling gradually became larger and heavier during development in order to carry a more lethal warhead. This is an example on its traversable test launcher.

Hs 297 to 8-117 or its company designation, Hs 117. The initial tests were unguided and were mainly intended to verify the performance of the boosters and sustainer engine. Other test missiles were dropped from aircraft without engines to test the flight control and guidance systems. The 8-117B had a revised arrangement of the sustainer fuel tanks and different mounting of the rocket boosters. The final trials version in the autumn of 1944, the 8-117C, had the warhead increased from 25 to 40kg, used the Colmar radio receiver, and was configured with a modular fuze pocket that could accept the Donag Kakadu, Siemens Marabu, or AEG Fuchs proximity fuzes. No specific proximity fuze had completed development, so the missile had to be prepared for any one of the several designs. The 8-117D was the same as the 8-117C but used the Walther HWK 109-729 engine. Speeds up to Mach 0.9 were reached during trials. In total, 79 flight tests were conducted through February 1945, 20 from aircraft and 59 from ground launchers. The full performance envelope could not be tested due to delays in engine development.

The first production version of the missile was the Hs 117A-1, based on the 8-117C test version. The standard production version was scheduled to be the Hs 117A-2, which had a revised spoiler/aileron system, tapered tail, modular compartment for either Straßburg, Colmar, or Brig radio receivers, and additional fittings for the optional Meise acoustic proximity fuze. Henschel was given a production contract for the Schmetterling in the summer of 1944. A second-generation version of the missile also appeared on the drawing boards, with the existing types now labeled Schmetterling S1 and the next generation as S2. There was a study of an improved supersonic configuration dubbed the Zitterrochen (electric ray) with a truncated delta wing configuration.

It had been hoped to start Schmetterling production by the end of 1944. However, many issues remained to be resolved. No actual live test

The Hs 117 Schmetterling had the characteristic asymmetric nose of Henschel antiaircraft missiles with a proximity fuze in the elongated nose and the propeller for an electrical generator on the opposite side. This Hs 117 on display at the Udvar-Hazy Center of the National Air and Space Museum lacks the two Schmidding rocket-assisted take-off boosters. (Author)

of the missile against an aircraft target had been attempted, nor was any proximity fuze ready for production. As of October 1944, the plan was to begin production in March 1945. One of the main bottlenecks was the supply of the rocket motors, with BMW reporting on January 21, 1945, that it could not deliver engines for at least another three months. The ambitious plan to produce 3,000 missiles monthly by late 1945 was trimmed back on March 13, 1945, to only 300 monthly due to extreme fuel shortages and the devastation in the aircraft industry caused by Allied bomber attacks. A total of 35 fire control stations were also on order. Since none of the types had been fully qualified in testing, four different configurations were ordered: 20 Burgund, 5 Franken, 5 Elsaß, and 5 Braband.

The Hs 117 Schmetterling used the Burgund radio-command guidance system, as shown in the postwar intelligence report. The Burgund Flakfolgegerät was a two-man fire direction post with the gunner on the right tracking the missile optically, and using a joystick to change the missile flight path via a radio command link. The operator on the left kept the post pointed at the missile.

The Burgund was the most likely fire control system for the Schmetterling in the short term, and was the closest to completion. It functioned in a similar manner to the Kehl-Strasburg system on the Hs 293 antiship missile, except for the fact that the ground-tracking elements were considerably more elaborate than the Kehl aircraft system due to the greater range and higher speed of the target. The fire control sequence began with an alert from a search radar at the launcher battalion's command post to the Flaksichgerät with each launch battery. Early plans assumed a single radar for every four fire units. The Flaksichgerät was an optical range-finder used to calculate the range, altitude, and direction of the target. Its data was passed via cable to the Flakfolgegerät, which was a two-man guidance station mounted on a semi-mobile Sd.Anh. 206 cruciform. The operator on the left kept the traversable frame pointed towards the missile and target, while the guidance operator on the right controlled the missile. The initial version of the control station used a Revi gun sight of the type used on German fighters, to track the missile and target. This was insufficient for use with targets at greater ranges and a telescopic sight was eventually added. The guidance operator controlled the missile using the usual joystick, much in the fashion of the earlier Hs 293 antiship missile. The fire control station was connected to a Funksender 203V transmitter antenna which served as the radio command link to the missile, sending course corrections to the Colmar receiver on the missile. As can be surmised from this description, this fire control system was limited to daylight, fair-weather conditions.

By March 15, 1945, when the Henschel plant was evacuated, a total of about 140 Hs 117 missiles had been assembled, of which 80 had been expended in tests, with the other 60 still missing key components. The development team was moved to the Harz mountain area in early 1945, and then to Bavaria in April 1945, with little opportunity to complete development of this missile system. The Schmetterling was the only German Flak missile close to service deployment in 1945.

Messerschmitt Enzian

The Enzian (Gentian herb) was Messerschmitt's contender for the subsonic Flak missile requirement and was led by Dr Hermann Wurster. It resembled a smaller, unmanned version of the Me 163 rocket fighter. The first version, called Flakrakete FR-1, completed preliminary design in June 1943, followed sequentially by FR-2, FR-3, etc. The first significant version was the FR-5, designed in October–November 1943. It was fitted with the Walter RI-203 rocket motor as a stop-gap because the planned Walter HWK 109-739 engine was delayed. The RI-203 was derived from a rocket-assist take-off unit for conventional aircraft and used hydrogen peroxide for the oxidant. The FR-6 was a late preproduction design from October 1943 to April 1944, mounting the Walter HWK 109-739 rocket.

The test versions of the Enzian Flak missile used a modified 88mm Flak gun mount with a 6m launch rail.

The first serial production type was the Enzian E-1, built in February 1944 on the basis of the FR-5 design. About 60 were built and it was the only version of the Enzian produced in quantity. This type entered trial launches from the island of Greifswalder Oie, located in the Baltic to the northeast of Peenemünde, starting in March 1944. The Enzian E-2 was a still-born design that attempted to incorporate more wooden construction into the design due to shortages of light metals in the German aviation industry.

The Enzian E-1 was fitted with four Rheinmetall-Borsig solid-fuel di-glycol booster rockets. This booster had first been developed for the experimental Feuerlilie test missile. The booster rockets proved to be a major source of frustrations during the initial flight trials of the Enzian. Each of the four boosters was attached to the missile fuselage with a pair of explosive bolts that were activated when the engine burned out. However, burn-times were erratic, so instead of all four boosters releasing from the fuselage simultaneously, they peeled off in a haphazard fashion, creating asymmetric thrust and drag that resulted in frequent crashes. The solution

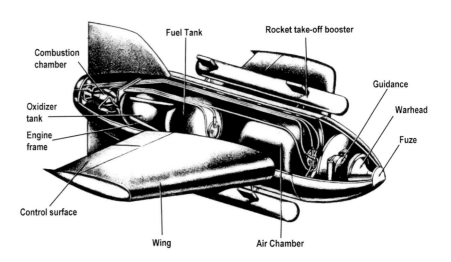

This cross-section drawing shows the layout of the Enzian. Other warhead configurations were under consideration.

was to add sensors linked to a single switch that detonated all the explosive bolts simultaneously. Only about a third of the 38 test launches were successful due to these propulsion problems.

The Walter HWK 109-739 encountered lingering design delays and only two units were ready for bench-test by early 1945. As an alternative, the Enzian E-3 was modified to use the VFK 613 A-1 engine developed by Dr Konrad for the Rheintochter 3. Although the E-3 design was complete, this version did not reach the test stage since no engines were available. The E-5 was a

An Enzian in its colorful test colors on display at RAF Museum Cosford. This museum has one of the most extensive collections of German World War II missiles. (Author)

supersonic version of the Enzian on the drawing boards in 1945, but was never built. Enzian guidance development was badly behind schedule and no guided test flights, except for basic flight controls, were conducted.

Due to shortages of light metals for aviation construction, the Enzian made extensive use of wooden construction for the wing and fuselage. The initial test missiles were assembled at the main Messerschmitt plant in Augsburg from late 1943 until early 1944. The Augsburg plant was a major target of the Allied bomber offensive in early 1944, so the Enzian program was transferred to the Holzbau Kissing KG plant in Sonthofen, near the border with Austria, which specialized in advanced wood construction techniques. The Enzian program was canceled on January 17, 1945, since it was seen as interfering with the Me 163 rocket fighter program. By the war's end, about 60 Enzians had been completed, of which 24 had been test launched and ten more expended in various ground tests. About 15 missiles were burned at the Sonthofen plant on April 25 to prevent their capture.

The Wasserfall

The Wasserfall supersonic Flak missile was the most sophisticated and expensive weapon of the four major Flak missile programs. The program was managed by the Army's Peenemünde development center under Gen Walter Dornberger, with Werner von Braun as the chief engineer. In May 1941, Dornberger instructed the design team to begin a preliminary study

A view of the first Wasserfall W1 test launch from Greifswalder Oie island on February 29, 1944. At this stage, the Wasserfall had relatively large mid-body wings which were soon recognized to be poorly suited for maneuvering against a moving target.

of a ground-based missile system capable of intercepting bombers at an altitude of 15,000–18,000m (about 50,000–60,000ft). The plan was to exploit the technology already in development for the A-4 (V-2) ballistic missile for the new system, tentatively called Vesuvius.

Three basic configurations were considered: the C1 single stage solid-fuel missile, the C2 liquid-fuel design, and the C3 two-stage missile. The designs closely resembled the A-4 ballistic missile since aerodynamic tests by this stage were well underway. However, the missile was significantly smaller than the A-4 since it only needed a 250kg (550lb) warhead

The C2 Wasserfall bore a strong family resemblance to the A-4/V-2 ballistic missile but was considerably smaller, with mid-fuselage wings and enlarged control surfaces on the tail. This is the standard W5 configuration after the mid-body wings had been reduced to a smaller trapezoidal shape.

and its maximum range was only 17km (10.5 miles) due to the limitations of existing guidance technology. Compared to the A-4 configuration, the missile had additional fins to permit maneuvering against its target.

The solid-fuel C1 was short-lived since it overlapped other programs such as Rheinmetall's Rheintochter program. As a result, the liquid-fuel C2 single stage missile became the focus of development. The first detailed design study by von Braun was completed on November 13, 1942. By this stage, Göring had changed his views in favor of Flak missiles, so the program was on firmer political grounds than before. In May 1943, the program name was changed from Vesuvius to Wasserfall (Waterfall).

The most substantial difference between the C2 Wasserfall and the A-4 was the decision to employ a rocket motor with hypergolic fuel. The A-4 used super-cold liquid oxygen as the oxidant. While very efficient, it was difficult to maintain the cryogenic conditions in the fuel tank for any period of time and the liquid oxygen tended to boil off. This was impractical in a Flak missile that might have to sit fully fueled on its launch pad for hours or days. Instead, von Braun proposed the use of nitric acid as the oxidant, with Visol (vinyl isobutyl ether) as the fuel. Hypergolic fuels were not as efficient as cryogenic fuels, but they were more practical for the Flak mission. A prototype engine was tested at Peenemünde in March 1943 and the preproduction engine was completed in July 1943. Due to the high cost and complexity of the turbo-pump to feed fuel into the A-4 engine, the Wasserfall used the simpler method of fuel injection by compressed air.

Early tests of the C2 with mid-body wings discovered that the additional control surfaces actually made the missile too stable, and therefore very difficult to maneuver. As a result, the wing design was changed from the large wings on the W1 version to a truncated pyramid design on the improved W5 version, which proved effective up to Mach 3. Like the A-4, the Wasserfall employed graphite vanes in the rocket efflux for steering. However, the maneuvering requirements led to changes in the tail design of the W5 with the addition of more extensive tail surfaces.

The superficial similarity between the A-4 and Wasserfall led to wildly optimistic predictions about its readiness for production. The initial schedule called for the completion of production drawings by April 15, 1944. A review of the program in January 1944 revealed that the program was far from mature, with matters such as guidance and engine design far from ready. Ideally,

THE C2 WASSERFALL FLAK MISSILE

The initial version of the C2 Wasserfall missile, the W1, was fitted with relatively large mid-body fins. Trials and wind-tunnel tests revealed that these made the missile too stable, and less able to maneuver against a target. The definitive test version of the C2 Wasserfall was the W5 version, which used smaller trapezoidal wings and enlarged tail surfaces. Both missiles were finished in a test pattern of black and white that made it easier for observers to determine the roll rate of the missile during ascent. The service version would have probably been finished in duller colors or in camouflage like the A-4/V-2.

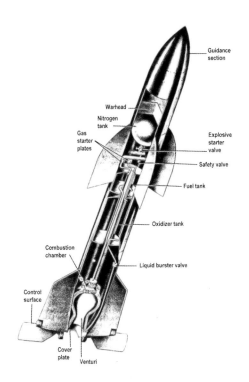

Guidance
section

Warhead

Nitrogen
tank

Gas
starter
plates

Explosive
starter
valve

Safety valve

Fuel tank

Oxidizer tank

Combustion
chamber

Liquid burster valve

Control
surface

Cover
plate

Venturi

This cross-sectional drawing
shows the basic internal layout
of the C2 Wasserfall.

the Wasserfall was supposed to be guided using radar tracking via a modified Würzburg-Reise radar, with the terminal phase being handled by some form of advanced guidance such as an infra-red seeker. These technologies were in their infancy, so there was the concession that the early version would use a form of the Kehl-Straßburg system of optical command guidance via radio such as the Burgund system. Speer's RBM armament ministry warned the Peenemünde staff that there was no large industrial capability to provide the volume of Visol fuel need for thousands of Wasserfall missiles, forcing the team to hunt for a new propellant and to adapt the engine to whatever fuel combination emerged.

Von Braun wanted to wait until late 1944 to begin launches with a refined Wasserfall design. The Luftwaffe was insistent that test flights begin by late 1943, given the urgency of the requirement. The first test launch was conducted from Greifswalder Oie island on February 29, 1944. The test missile had a gyroscopic platform for basic flight control and roll stabilization, but no actual guidance system since none had yet been developed. In the event, the flight control system failed shortly after launch and the first Wasserfall launch failed. A second test launch on March 8 also failed. Siemens delivered an improved gyro flight control system for the third launch on May 12, but the missile failed 22 seconds into the launch, probably due to the weak servomotors used to activate the control surfaces. The fourth launch on June 8 ended in spectacular failure when the explosive bolts holding the missile to the launch-stand failed to operate, with the missile lifting off with the launch-stand attached, crashing near the pad 9 seconds later. A fifth test flight in July ended prematurely when the engine exploded.

The immaturity of the Wasserfall design led the design team to modify an A-4 missile with a radio-command guidance system, and it was test-launched on June 13. The operator was able to control the missile via a joystick for about 30 seconds, at which point the missile disappeared from view. It ended up exploding over Sweden. The Swedish government later passed on missile components to British agents, leading to the mistaken assessment that the A-4/V-2 used radio guidance.

The pace of development and testing accelerated in the second half of 1944 as the technical problems were gradually ironed out. By the end of 1944, a total of 25 test missiles had been launched, with ten failures. Many of the test launches failed to reach the intended range. The problem was that the fuel sloshed around in the tank, allowing the pressurized nitrogen used for fuel injection to inadvertently leak into the combustion chamber, leading to erratic combustion. This problem had still not been solved by early 1945.

By August 1944, nearly a quarter of the development staff at Peenemünde was working on the Wasserfall program. Plans in the summer of 1944 called for production of the first 1,000 missiles using the Burgund joystick radio-control. The following 5,000 missiles would use a Telefunken radar guidance system still under development, supplemented by the Karussel infrared seeker for terminal guidance.

The Burgund joystick command system proved practical, although it was easily interrupted by cloud cover. Without a mature radar guidance system, it was recognized that the range requirement was illusory since the weapon could only be used in clear weather. Another version, called the Wasserfall W10, was proposed, that was smaller and more economical. Its shorter range was more compatible to the restrictions of the early guidance system.

The initial radar guidance system, codenamed Elsaß (Alsace), used a Rheingold radar to track the location of the Wasserfall missile in conjunction with a Rüse transponder on the missile after launch. In the meantime, the enemy bomber was tracked by a Mannheim-Reise radar. The data from the two radars were fed to a fire direction calculator, passed on to a fire control crew, and the course corrections sent to the missile via a radio link. The Elsaß system was months or years away from production at the war's end. Not only was the Wasserfall guidance system not ready, but Speer's armament ministry warned Peenemünde that they could not guarantee an adequate supply of graphite needed for the engine guidance vanes.

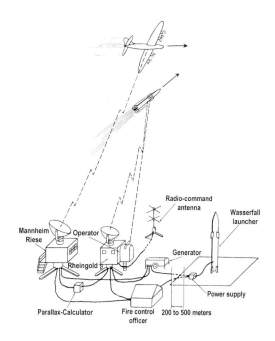

The Luftwaffe intended to use the Elsaß guidance system for the Wasserfall, but it was far from mature when the war ended. This system would have used a Mannheim-Riese radar to track the target and a Rheingold radar to track the missile, with radio command signals to guide the missile to the target, as shown in this postwar intelligence report.

Reassessing the Flak Missile Program: Late 1944–Early 1945

By the autumn of 1944, Allied heavy bombers were systematically destroying the German armaments industry, transportation network, and synthetic fuel plants. The Luftwaffe's fighter force had been decimated and its Flak artillery force degraded by Allied electronic countermeasures. It was only a matter of time before the German war economy completely collapsed. This added a much greater sense of urgency to the Flak missile program. At the same time, it was becoming increasingly clear that the four major missile programs were far from ready.

Aside from the enormous development challenges, it was by no means clear that Germany had the resources necessary to produce such complex weapons in sufficient quantity. Rudolf Brée, the head of the guided weapons development department of the RLM/GL, dismissed the Flak missile concept in a 1944 report as "an immensely expensive consumer item for the 'filthy rich.'" Proponents of the Flak missiles such as Oberstleutnant Friedrich Halder retorted that it would be less expensive than current Flak guns. For example, in 1944 it took an average of 16,000 rounds of 88mm Flak to down an Allied heavy bomber, which consumed 85 tonnes of nitric acid to manufacture the ammunition propellant. He claimed that two Wasserfall missiles had the same probability of downing a bomber and consumed only 3 tonnes of nitric acid for propulsion. This seemed like an impressive argument in favor of Flak missiles except for the fact that the Wasserfall had not conducted a single successful guided flight up to this point.

By the autumn of 1944, the Amtsgruppe für Flakentwicklung of the RLM/GL completed an assessment of the major programs with an aim to winnow the program down to the most advanced that had some reasonable likelihood of reaching the production stage by 1945. They recommended that only three programs be continued. The Taifun rocket was accepted simply because it was so elementary and required such modest industrial resources; production was allotted to the Benteler Werke in Bielefeld. The only two guided missile programs that were approved were the Henschel Schmetterling and the Wasserfall; Enzian and Rheintochter were to be ended. The Henschel program was by far the most mature of all the German Flak missiles, in no small measure due to the experience of the firm since 1941 with a wide range of missiles. The Wasserfall was also approved in spite of its staggering technological problems, since it represented the only hope to deal with future threats such as the Boeing B-29 Superfortress bomber that was expected to appear over Germany in 1945.

The accompanying chart shows the plans for the serial manufacture of Flak missiles as of October 20, 1944. As can be seen, Wasserfall production was seven months behind Schmetterling production.

Planned Flak Missile Production, March 1945–March 1946

	Mar '45	Apr	May	June	July	Aug	Sep	Oct	Nov	Dec	Jan '46	Feb	Mar
Schmetterling Missiles	150	250	400	700	1,100	1,500	2,000	2,500	3,000	3,000	3,000	3,000	3,000
Schmetterling Basic Fire unit	1	2	2	4	4	5	6	7	7	6	5	4	3
Fire unit with FuMG 44 radar	–	–	1	1	1	1	1	–	–	–	–	–	–
Fire Unit with FuMG 39 Ansbach radar	–	–	–	–	3	3	4	4	4	4	4	4	4
Wasserfall missiles	–	–	–	–	–	–	–	50	100	200	400	600	900
Wasserfall fire units	–	–	–	–	–	–	–	–	1	2	4	5	7

On October 30, 1944, Luftwaffe chief Hermann Göring, accompanied by armaments czar Albert Speer, were given a demonstration of the various Flak missiles at Peenemünde. This had been arranged by Axthelm, von Renz, and other Flak missile proponents in the hope of securing stronger political commitment to the program in Berlin. The demonstration backfired. The drug-addled Göring was very impressed by the fiery ascent of the Enzian missile with its four booster rockets. Without any technical appreciation of the program's problems, he decided to continue the Enzian even though the program was hopelessly behind schedule. Speer was not impressed with the progress on the Wasserfall and was concerned by its growing costs. He recommended that it be downgraded to a technology demonstration effort and not included in the Luftwaffe's 1945 production plans. In the event, these recommendations were largely evaded, with the Luftwaffe depending on Hitler's November 4, 1944, decree that Flak programs should be accelerated to stun the Allies with fear of the "hell of German Flak fire."

To further add to the chaos, in November 1944, Hitler sanctioned the take-over of the Luftwaffe Flak missile program by SS-Gruppenführer Hans Kammler as the overall missile czar. He had already taken over control of the Mittelwerk missile plant at Nordhausen and had been given tactical command of the V-1 and V-2 units.

A number of configurations were under study for Wasserfall launch bases, including this concept with a circular preparation building with multiple rail-tracks to move the missiles into position for launching.

Plans to begin the production of Flak missiles in 1945 were disrupted by Germany's declining fortunes. When the Red Army unleashed its Vistula-Oder offensive on January 12, 1945, it seemed only a matter of weeks before Peenemünde fell into Soviet hands. The design staffs began evacuating to Bavaria. On January 13, 1945, Speer created the Arbeitsstab Dornberger (Working Staff Dornberger) to manage the various missile programs, including the Flak missiles. This gave Dornberger special powers regarding production decisions. Plans to manufacture the Wasserfall at the Elektromechanische Werke GmbH, the Peenemünde manufacturing plant, were no longer feasible. On January 28, all Flak missile development efforts were ordered to move to the Mittelwerk tunnels at Nordhausen in central Germany, under Kammler's control. The transfer began on February 17, 1945.

Seeing Speer's actions as a threat to his powers, Kammler went to Göring on January 26 and demanded that Dornberger's staff be subordinated to him under his new role as Commissar for Brechung des Luftterrors (Breaking the Air Terror). With Himmler's backing, Kammler on February 6, 1945 announced a cancelation of the R-Programm for Flak missiles except for those nearing the production phase, namely Schmetterling and Wasserfall. The Rheintochter and Enzian were specifically identified for termination. By this time, the decision had little consequence given the chaotic conditions in Germany. With the evacuation of Peenemünde in February 1945, further missile development and testing collapsed. The manufacturing facilities at the Mittelwerke in Nordhausen were already committed to V-1 and V-2 production, and it would take months to gear up for production of Schmetterling and Wasserfall.

An assessment by the Reichsforschungsrat (Reich Research Council) of February 26, 1945, noted that the Flak missile needed to master four technologies: rocket propulsion, aerodynamic design, missile guidance, and proximity fuze. The report concluded that engine and missile designs were relatively mature, but that neither guidance nor proximity fuzes were within immediate grasp. The report concluded that "none of the four missile programs have a chance of success within the next six months." The hope for a more advanced guided missile relying on more sophisticated guidance than a simple joystick control would not be ready "until the next war." At the war's end, no Flak missile had conducted test interception of an actual target aircraft and none of the missiles had been tested with a functioning proximity fuze.

After the war, Speer recognized that the decision to focus on the V-weapons instead of the Flak missiles was a fundamental mistake:

> To this day I think that the Flak missile, in conjunction with the jet fighters, would have repulsed the Western Allies' air offensive against our industry from the spring of 1944 onwards. Instead, gigantic effort and expense went into the development and manufacture of long-range missiles that proved to be, when they were ready in the autumn of 1944, an almost total failure. Those missiles, which were our pride and for a time my favorite armaments project, proved to be nothing but a mistaken investment. Furthermore, they were one of the reasons that we lost the defensive war in the air.

Flak Missiles Technical Data

	Schmetterling	Rheintochter III	Enzian	Wasserfall W5
Length (m/ft)	4.29/14	5.0/16.5	3.65/12	7.85/25.7
Diameter (m/ft)	0.35/1.1	0.5/1.6	0.89/2.9	0.88/2.9
Wingspan (m/ft)	2.0/6.6	2.2/7.5	4.2/13.8	1.98/6.5
Launch weight (kg/lb)	445/980	1,565/3,450	1,970/4,350	3,540/7,800
Warhead (kg/lb)	25/55	160/350	475/1,050	250/550
Rocket motor	HWK 109-729	VFK 613	VFK 613 A-01	EMW R IX
Speed (km/h/mph)	860/535	1,200/750	900/560	2,375/1,475
Max range (m/ft)	16,000/52,500	18,750/61,500	25,000/82,000	48,000/157,500
Max altitude (m/ft)	10,650/35,000	15,000/49,250	14,500/47,500	20,000/65,500

AIR-TO-AIR MISSILES

By 1943, existing aircraft machine guns and cannon were insufficiently lethal against Allied heavy bombers. Various types of unguided rockets were developed for bomber attack, but these had poor accuracy. In June 1943, the Luftwaffe began an air-launched guided missile program as a rival to the Flak missile program. Dr Max Kramer and Ruhrstahl proposed a wire-guided missile based on previous work on a wire-guided version of the Fritz-X. The small 8-344 missile, later called X-4, was powered by a BMW 109-538 liquid-rocket engine. The guidance system used a pair of trailing wires derived from the Dortmund/Duisberg system developed earlier for antiship missiles. After launch, the pilot tracked the missile by following its tracer flare, and steered the missile using a

Kramer's X-4 air-to-air missile used wire guidance, with the wire spools contained in the tear-drop fairing on the wing-tips. This example is preserved at the Museum of the US Air Force at Wright-Patterson AFB. (Author)

Knirps joystick control. The warhead was triggered either by a Rheinmetall-Borsig Kranich acoustic proximity fuze or an impact fuze.

Test launches began in September 1944 and about 160 test missiles were built. The tests showed that the missile was technically sound except that the liquid-rocket fuel was dangerous. An alternative solid-rocket motor entered development at WASAG, which already manufactured rocket motors for the Panzerschreck antitank rocket. There were plans to manufacture the missile in substantial quantities, reaching 5,000 monthly.

There was some skepticism about the tactical merits of the missile since the parent aircraft could not maneuver during the 10-plus seconds of missile flight, making it vulnerable to Allied escort fighters. Originally, it was planned to operate the X-4 from the FW 190 with two missiles per aircraft. It was later decided to shift this to the faster and less-vulnerable Me 262 jet fighter with four missiles per plane. In the event, the X-4 was canceled by Kammler's edict of February 6, 1945, that shut down many of the German missile programs.

After the war, the French air force manufactured a copy at Arsenal de l'aéronautique as the AA 10. About 200 were built for tests that lasted from 1947 to February 1950. The French concluded that the liquid-rocket engine was a maintenance hazard since it had to be fueled with dangerous nitric acid oxidant prior to flight. The guidance system offered only marginal accuracy, so the air force waited for the advent of a solid-fuel, command-guided missile, the Nord AA.20, in 1956 before starting mass production.

This view of a disassembled X-4 provides a good view of the BMW 109-538 liquid-fuel rocket engine. The box immediately in front of the engine is the missile battery, while the conical gyro/flight control unit has been removed from its position behind the wing root and can be seen below the tail assembly.

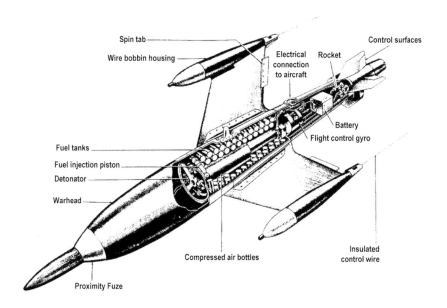

A cut-away illustration of the X-4 air-to-air missile.

ABOVE LEFT
A Henschel Hs 298V-1 on an underwing rail launcher during its initial trials at Peenemünde. About 100 of this version were built but it proved unsatisfactory, forcing a substantial redesign.

ABOVE RIGHT
The Henschel HS 298V-2 reversed the position of the proximity fuze and generator in the nose and used a redesigned tail. It was powered by a Schmidding 109-543 solid rocket. This example is on display at the US National Air and Space Museum's Udvar-Hazy branch in Chantilly, Virginia. (Author)

Air-to-Air Missiles

	X-4	HS 298
Length (m/ft)	1.9/6.5	2.05/6.7
Wingspan (m/ft)	0.57/1.9	0.62/2.04
Warhead weight (kg/lb)	20/44	50/110
Guidance	wire	radio-command
Proximity fuze	Kranich acoustic	Fuchs radio-frequency
Launch weight (kg/lb)	60/132	120/265
Engine	BMW 109-538	Schmidding 109-543
Rocket fuel	liquid	solid
Thrust (avg kg/lb)	30/70	50/110
Speed (km/h/mph)	900/560	870/540
Max range (m/ft)	2,800/9,200	4,500/14,750

In parallel to the X-4, Dr Wagner at Henschel developed the Hs 298 air-to-air missile. This resembled a miniature version of the Schmetterling Flak missile, and used the proven Kehl-Straßburg radio command system. Compared to the X-4, it was nearly double the weight but had a much larger warhead and greater range. The tactical concept was to launch the Hs 298 from a heavy fighter such as the Do 217 or Ju 88 from a long stand-off range outside the enemy bomber's machine gun. It could also be employed by heavy night fighters. The Hs 298 was intended to be operated by a two-man team using the Fevi gyro-stabilized sighting system.

The first test on December 22, 1944, was conducted from a Ju 88C. This did not go well; one missile blew up shortly after launch, one nose-dived and crashed, and one was stuck on the launch rail. The Do 217 could carry five missiles – with one under each wing and three under the fuselage. There were also plans to launch it from the Fw 190, though presumably this would have required a different optical tracking system. The Luftwaffe ordered an

 X-4 AIR-TO-AIR MISSILE

Although the X-4 was tested from the Fw 190 fighter, the plan was to deploy it primarily on the Me 262 jet fighter. Each fighter was to carry four missiles on underwing pylons. The missile was controlled optically by the pilot using a Knirps joystick controller, sending course corrections via the wire link. Absolute precision was not required since it was anticipated that the missile would be fitted with a Kranich acoustic proximity fuze. French tests of an X-4 copy after the war found the performance of the missile extremely disappointing. Controlling a small missile from a fast-moving fighter using a simple joystick control was a very inadequate form of guidance.

Not convinced that a simple joystick would prove precise enough in the dynamic conditions of air warfare, the Hs 298 could be controlled from a heavy fighter such as a Ju 188 using the Fevi gyro-stabilized guidance system. One of the operators tracked the target while the other controlled the missile.

initial production batch of 2,500 in April 1944, followed by a contract for 2,000 more in July 1944. The plan was to manufacture the Hs 298 at a rate of 2,000 monthly, but the program was canceled along with many efforts on February 6, 1945. It's worth noting that Henschel also tested an air-to-air version of the Schmetterling, the Hs 117H, and an air-to-air version of the Hs 293 antiship missile, the Hs 293H.

ANTITANK MISSILES

In 1941, BMW made an unsolicited proposal to the Heereswaffenamt (HWA: Army weapons bureau) for the development of a guided antitank missile. It was rejected due to its cost. By 1943, HWA had reconsidered this issue and awarded three development contracts for a guided antitank missile with a range of 1km.

Dr Kramer at Ruhrstahl proposed the X-7 Rotkäppchen (Little Red Riding Hood) wire-guided antitank missile based on the technology developed for the X-4 air-to-air missile. The missile, officially designated as the 8-347, was a flying-wing design with a small fuselage containing a shaped-charge warhead and a WASAG 109-506 solid-rocket, two-stage motor. There were pods at the extremities of both wings containing the wire spools used by the guidance system. The missile operator used a joystick connected to a FuG 510 Düsseldorf transmitter which passed the commands along the wire to the FuG 238 Detmold receiver on the missile. The missile flight control was based on a simple gyro for roll stabilization and the missile rotated at a rate of one revolution per second for stability. Tests began on September 21, 1944, at the Sennelager training base. Of the seven test missiles, four crashed within 25m of the launcher, two engines blew up, and one missile hit a target at 500m. Subsequent tests encountered frequent crashes, and eventually the roll rate was increased to two revolutions per second, with a stub tail added to improve the missile's stability. Testing was later conducted at Peenemünde-West due to better telemetry equipment. Small-scale production of about 300 missiles began at the Heber Mechanischen Werke in Neubrandenburg. Plans to manufacture 20,000 of these missiles monthly collapsed in early 1945. Three launcher concepts were studied but none was selected. Lacking launchers and control equipment, the missiles were never issued to the troops and so were stored in the B3a tunnel complex in Himmelberg and the Aladinhöhle cave in Thuringia. Curiously enough, none of these were captured by the Allies, except for a single example of a badly damaged test missile.

No examples of the Kramer X 7 Rotkäppchen appear to have survived, but this illustration shows its basic configuration. To stabilize the missile in flight, it rolled at a rate of two revolutions per second, with the engine angled slightly off the center-line to contribute to this motion. (Author)

At least two other missiles were developed in this time frame. Dr Kluge at AEG offered the Rumpelstilzchen. About 100 of these were built for tests, but no details have survived. BMW offered a design powered by two solid-rocket motors. Since these were not yet available, the BMW test missiles used a BMW 109-448 liquid-rocket motor. The design resembled a flying wing and used wire guidance. Initial tests were conducted near the Dachauer Moor. Although the tests were successful, there does not appear to have been any significant production due to the collapse of German industry at this point in the war. There were also a number of paper designs that did not reach the test stage.

Postwar Influence of the German Missile Programs

After the war, Allied armies went to great lengths to collect wartime German technology. The French navy built a small number of Hs 293 copies at the Brest Arsenal in the late 1940s, but moved on to more sophisticated designs. Likewise, the Soviet Union planned to build copies of the Hs 293 at Plant No. 272 in Leningrad in 1947–48, but reconsidered after testing captured examples from Tu-2D bombers. Curiously enough, the Argentinian state factory FMA built a small number of Hs 293 copies in 1952–56, launched from Lancaster bombers. Czechoslovakia also assembled some Hs 293s after the war from captured parts.

Although German Flak missiles drew intense scrutiny in the US and Britain, the lack of a proven guidance system or a functioning proximity fuze lessened interest in the programs. The US built an enlarged version of the Wasserfall as the Hermes, but this was a technology demonstrator for liquid-rocket propulsion and not intended as an air defense missile. The Soviet Union reverse engineered the Schmetterling as the R-102/R-117 and the Wasserfall as the R-101 but did not put them into service use. The first Soviet air defense missile, the S-25 (SA-1 Guild), traced its propulsion lineage to the Wasserfall; it used a completely different guidance system. Curiously, the Wasserfall rocket engine also led to the liquid-fuel engine of the Scud ballistic missile family.

France showed interest in the wire-guided antitank missiles that led to postwar designs such as the Entac and SS-10. The French air force also tested the X-4 antiaircraft missile in 1947 but found it to be unsatisfactory. German engineers working for French firms established a tradition of Franco-German cooperation in missile development, leading to such firms as Euromissile and today's MBDA.

FURTHER READING

Government Reports

Benecke, T. and Quick, A. W., *History of German Guided Missiles Development*, Advisory Group for Aeronautical Research and Development, NATO (1957)

Bramble, Robert, *Development of German Antiaircraft Guided Missile Wasserfall*, HQ, Air Materiel Command, Wright Patterson AFB (1948)

Hutcheon, I. C., *German Non-Guided Flak Rocket-Taifun*, Armaments Design Dept Technical Report 6/46, Fort Halstead, Kent (June 1946)

Von Renz, Gen der Flakartillerie Otto W., *The Development of German Anti-Aircraft Weapons and Equipment of All Types up to 1945*, US Air Force Historical Study 194 (1958)

Tsien, H. S., *et al.*, "Historical Notes on German Guided Missile Development", in *Technical Intelligence Supplement*: AAF Scientific Advisory Group, HQ Air Materiel Command, Wright Field (May 1946)

n.a., *BV 246 Glide Bomb*, US Naval Technical Mission in Europe, Report No. 7-45 (June 1945)

n.a., *German Air-to-Air Missile X-4*, US Naval Technical Mission in Europe, Report No. 30-45, (May 1945)

n.a., *Handbook on Guided Missiles of Germany and Japan*, Military Intelligence Division, US War Department (1946)

n.a., *Survey of German Activities in the Field of Guided Missiles*, US Naval Technical Mission in Europe, Report No. 237-45 (August 1945)

Books

Benecke, Theodor, *et al.*, *Die deutsche Luftfahrt: Flugkörper und Lenkraketen*, Bernard & Graefe, Koblenz (1987)

Bollinger, Martin, *Warriors and Wizards: The Development and Defeat of Radio-Controlled Glide Bombs of the Third Reich*, Naval Institute Press, Annapolis (2010)

Boog, Horst, *et al.*, *Germany and the Second World War, Vol. VII: The Strategic Air War in Europe and the War in the West and East Asia 1943–45*, Oxford University Press, Oxford (2006)

De Maeseneer, Guido, *Peenemünde: The Extraordinary Story of Hitler's Secret Weapons*, AJ Publishing, Vancouver (2001)

Griehl, Manfred, *Deutsche Flakraketen bis 1945*, Podzun-Pallas, Wolfersheim-Berstadt (2002)

Griehl, Manfred, *Deutsche Raketen und Lenkwaffen bis 1945*, Motorbuch, Stuttgart (2015)

Griehl, Manfred, *Do 217-317-417: An Operational Record*, Airlife, Shrewsbury (1991)

Hahn, Fritz, *Waffen und Geheimwaffen des deutschen Heeres 1933–1945, Band 2*, Bernard & Graefe, Koblenz (1987)

Klee, Ernst and Merk, Otto, *The Birth of the Missile: The Secrets of Peenemünde*, George Harrap, London (1965)

Materna, Horst, *Die Geschichte der Henschel Flugzeug-Werke in Schönefeld bei Berlin 1933–45*, Rockstuhl, Bad Langensalza (2016)

Neufeld, Michael, *The Rocket and the Reich: Peenemünde and the Coming of the Ballistic Missile Era*, Free Press, New York (1995)

Nowarra, Heinz, *Deutsche Flugkörper*, Podzun-Pallas, Friedberg (1987)

Roba, Jean-Louis, *Kampfgeschwader 100: L'escadre au Drakkar*, LELA, Le Vigen (2015)

Schabel, Ralf, *Die Illusion der Wunderwaffen: Düsenflugzeuge und Flakabwehrraketen in der Rüstungspolitik des Dritten Reiches*, Oldenbourg, Munich (1994)

Stüwe, Botho, *Peenemünde West: Die Erprobungsstelle der Luftwaffe für geheime Fernlenkwaffen und deren Entwicklungsgeschichte*, Bechtle, Esslingen (1995)

Trenkle, Fritz, *Die deutschen Funklenkverfahren bis 1945*, AEG, Ulm (1986)

Westermann, Edward, *Flak: German Anti-Aircraft Defenses 1914–1945*, University Press of Kansas, Lawrence (2001)

Wolf, William, *Fritz X: The Ruhrstahl SD 1400X Guided Missile*, Merriam Press, Bennington (1988)

Wolf, William, *Henschel Hs 293 Guided Missile*, Merriam Press, Bennington (1988)

Articles

Ludwig, Karl-Heinz, "Die deutschen Flakraketen im Zweiten Weltkrieg" in *Militärgeschichtliche Zeitung*, 1(1969), pp.87–100

Pawlas, Karl, "Die Flakrakete Hs 117 Schmetterling" in *Waffen Revue*, Parts 1–7, Vol. 73 (II/1989) – Vol. 79 (IV/1990)

Rimell, Ray, "The Kaiser's Guided Missiles" in *Aeroplane* (September 2008), pp.30–34

Saxon, Timothy, "Kehl: The German Use of Guided Weapons against Naval Targets 1943–44" in *Defence Studies*, Vol. 3 No. 1 (Spring 2003), pp.1–16

Sollinger, Gunther, "The Development of Unmanned Aerial Vehicles in Germany 1914–1918" in *Scientific Journal of the Riga Technical University*, Vol. 10 (2010), pp.24–30

INDEX